D1272410

Escaping from Enemies

Paul Bennett

Thomson Learning
New York

Nature's Secrets

Cover: A gemsbox oryx runs for its life in Namibia, Africa.

Title page: The anemone's stinging tentacles help to protect the hermit crab from its enemies.

Contents page: Ants will protect aphids and in return the aphids give out a sweet liquid the ants like to eat.

First published in the
United States in 1995 by
Thomson Learning
115 Fifth Avenue
New York, NY 10003

Published simultaneously in Great Britain by
Wayland (Publishers) Ltd.

Library of Congress Cataloging-in-Publication Data
Bennett, Paul, 1954–
Escaping from enemies / Paul Bennett.
p. cm.—(Nature's secrets)
Includes bibliographical references (p.) and index.
ISBN 1-56847-358-3
1. Animal defenses—Juvenile literature.
[1. Animal defenses.] I. Title. II. Series: Bennett, Paul, 1954– Nature's secrets.
QL759.B45 1994
591.57—dc20 94-24313

Printed in Italy

Picture acknowledgments
The publishers would like to thank the following for allowing their photographs to be reproduced in this book: Biofotos 13 (bottom/Andrew Henley); Bruce Coleman Limited 22 (M.P.L. Fogden), 4, 16/top (Jane Burton), 5 (bottom/Hans Reinhard), 6, 12 (Jen & Des Bartlett), 13 (top/Gerarld Cubitt), 14, 21/top (Leonard Lee Rue), 15 (inset/Gary Retherford), 17 (left/Erwin & Peggy Bauer), 17 (right/Wayne Lankinen), 18 (bottom/Dennis Green), 20 (George McCarthy), 25 (top/Fred Bruemmer), 28 (Francisco Futil); Natural History Photographic Agency *cover* (PeterJohns), *contents page*, 25/bottom (Andy Callow), 5 (center/Nigel Dennis), 5 (top/R. Kirchner), 7, 9, 15/top (Stephen Dalton), 8 (bottom/Laurie Campbell), 11 (top/Joe Blossom), 16 (bottom/Jeff Goodman), 23 (A.N.T.), 24 (Anthony Bannister), 29 (both/David Heuclin); Oxford Scientific Films Ltd *title page*, 8/top, 19/bottom (G.I. Bernard), 10 (main/Charles Palek and inset/Sean Morris), 11 (bottom/Zig Leszczynski), 15/main, 17, 21 (bottom/James H. Robinson), 18 (top/Renee Lynn), 19 (top/W. Gregory Brown), 26 (Fred Bavendam), 27 (David Fleetham).

Contents

Amazing defenses

For most animals staying alive is a constant battle. They must find food to eat, produce and raise young, and defend themselves when attacked by a hungry predator. To survive, animals have developed an incredible number of ways to avoid becoming a tasty meal.

The mole is a champion burrower. It lives, sleeps, feeds, and breeds underground, where it is safe from its enemies. ▽

△ This green lizard has lost its tail to a predator. The shed tail often twitches for several minutes, which confuses the enemy and gives the lizard a chance to escape. In time, a new tail will grow to take the place of the lost tail.

The American avocet lures a predator away from its nest by pretending to have a broken wing. It will fly back to the nest when danger has passed. ▷

The porcupine frightens off an attacker by sticking out its sharp, barbed quills. ▷ If touched, the quills can hook into the attacker's flesh, causing a painful wound. Threats are an animal's way of showing off its weapons to a predator.

It is difficult to spot an animal whose body is transparent. A predator may only see the water plants behind these glassfish. ▷

The getaway

A great many animals flee from their enemies as fast as possible in any way they can to get away.

△ The graceful springbok's large eyes and ears tell it that an enemy is nearby. With its long legs it can outrun a charging lioness.

△ A frightened basilisk lizard runs on water! It dropped from its riverside tree and is now dashing for safety. As it loses speed, it will sink into the water and swim away.

A grasshopper springs away from an enemy. It jumps by bending its long, slender hind legs and then suddenly straightening them. ▽

△ A queen scallop escapes from the clutches of a common starfish by squirting out a jet of water.

△ A moth flees from a hungry bat. The picture shows how the moth dives unexpectedly as the bat closes in for the kill.

◁ A cloud of snow geese take to the air. Frightened birds will fly away at the first sign of danger.

Retreat into cover

Some animals cannot run away. Instead they dive into a hole in the ground where a hungry attacker cannot follow. Others carry their homes with them and retreat into cover at the first sign of danger.

◁ The prairie dog has short legs and cannot run fast for long. Instead, it rushes to the safety of its burrow. Prairie dogs live in colonies and make a warning bark when an enemy is seen.

The trapdoor spider emerges from its burrow just far enough to catch insects. ▷

△ Box turtles escape by drawing their soft, vulnerable bodies into their armored shells. They seal themselves inside by folding the lower shells across the openings.

Lady crabs bury themselves in sand to hide from enemies. Often crabs dig in backwards so that they can keep a lookout for enemies as they dig. ▷

Rolling into a ball

To make a quick kill, most predators aim for the soft, vulnerable underparts or head of a victim. To protect themselves from this type of attack, some animals have sharp spines or horny plates on their backs and they can roll into a ball when an enemy comes near.

◁ The pangolin's armor of horny scales fit together so well that it can roll up into a ball. It will also squirt a foul-smelling fluid from a gland near its tail when in danger.

△ Having rolled itself into a near-perfect ball, the pill bug is well shielded against attack.

Like the porcupine, the echidna of Australia has a coat of sharp spines. It curls into a ball shape when in danger. It will peek out of its prickly protection when it thinks the coast is clear. ▷

Camouflage

Even though they are out in the open, many animals go unnoticed and unharmed because they are colored like their surroundings. They may have bodies that look like parts of plants too. Small, weak, and defenseless creatures are often well camouflaged and remain still until danger passes.

△ A four-day-old white-tailed deer sits motionless in its woodland home. The fawn's spotted brown coat makes it hard to see because the colors are like those of its surroundings.

The bulging eyes of the red-eyed tree frog make it very noticeable. ▷

But when it lies flat and closes its eyes, it is difficult to spot among the leaves. ▽

◁ Stick insects blend in wonderfully with the plants on which they live. This one looks like a leaf and will rock gently from side to side to mimic a leaf swaying in the breeze.

There is something hiding in the gravel on the sea floor. If you look carefully you will find a well camouflaged fish called a plaice. ▽

△ The willow ptarmigan has colors that make it blend in with its surroundings.

△ In winter, it even changes its plumage to match the snow.

The graceful chameleon can change its color to help it blend in with the background. ▷

Warning signals

Many animals are brightly colored as a warning to other animals that they are not good to eat. They do not need to hide from enemies.

◁ The skunk's bold black-and-white pattern warns predators not to attack. Instead of running or hiding in the face of danger, a skunk will spray a foul-smelling liquid from glands near its tail. It can squirt at an attacker up to ten feet away.

Many butterflies and moths make an unpleasant meal. The tiger moth's stunning scarlet and black wings warn that it is not good to eat. ▷

△ The beautiful lionfish can be found on coral reefs, but take care not to touch the poisonous spines on its back.

"Eat me at your peril" is the message this fire-bellied toad is sending to a hungry snake. The toad is lying on its back to show off its orange warning colors. ▽

Tricking the enemy

Some animals can fool an attacker into believing that they are dangerous. This is called bluffing. Bluffs are designed to startle or frighten an attacker into confusion or retreat.

△ When it sees a snake, the common toad puffs itself up with air and raises its body off the ground. Hopefully, the snake will think the toad is too big to swallow and leave it alone.

△ When challenged, the opossum will lie limp on the ground and will stay that way even if it is picked up! The phrase "playing possum" is used when someone plays dead to fool an enemy.

Eyespots can make a predator fear it is being attacked by one of its own enemies. A hungry bird may think this moth's false eyes are an owl's eyes and so fly away in haste. ▷

Some harmless animals cleverly mimic dangerous ones and so gain protection. The coral snake (left) is highly poisonous. Although the king snake (below) has markings very similar to a coral snake's, it is, in fact, harmless.

The spectacular Australian frill-necked lizard stands with its mouth wide open and its neck frill of skin erect to scare an enemy. It will bob its head and lash its tail back and forth too. ▷

Safety in numbers

An animal that lives on its own may have several defenses to outwit an enemy, but others benefit by living with others of their own kind or with another type of animal.

△ Meerkats are constantly on the look-out for danger. A meerkat will stand on sentry duty while the rest of the colony looks for food.

△ A group of musk oxen stand in a defensive circle to protect the young in the middle. The pointed horns will make a predator think twice about attacking.

Ants have strong, biting jaws and some have stings. They are ideal creatures to offer protection to other types of insects. These defenseless aphids give out a fluid the ants enjoy. In return the ants defend the aphids from predators. ▷

◁ A school of striped catfish will scatter in all directions when threatened. This makes it difficult for a predator to choose which fish to attack.

△ The clownfish retreats to the protection of a sea anemone. Animals that would attack the clownfish keep their distance for fear of being stung by the anemone's tentacles.

Fighting the enemy

If all else fails, an animal may defend itself by fighting its enemy. If the predator receives a painful wound, it may give up the attack and the animal will survive to live another day.

△ A hungry coyote rears back from the sharp, chisel-edged teeth of an annoyed beaver. The coyote crept up while the beaver was felling a tree.

△ This is a fight to the death. An Indian cobra prepares to strike at one of its enemies, a gray mongoose, which is famous for its ability to kill deadly snakes. It will leap out of range with remarkable speed.

Crayfish like to hide from their enemies, but if they cannot retreat into a hole in the riverbank, they will use their powerful claws to give a painful nip. △

Glossary

Breeds Mates and rears young.

Burrow A passage dug in the ground by an animal.

Colonies Groups of animals of the same kind living together.

Confuses Puzzles.

Erect To stand straight up.

Flee To fly, run, or swim away from an enemy.

Gland A part of the body that stores useful substances or gets rid of unwanted substances.

Lures Leads away.

Mimic To imitate.

Offspring The young of animals.

Outwit To defeat by cleverness.

Plumage The feathers of a bird.

Predator An animal that kills other animals for food.

Quills Sharp spines.

Retreat To draw back from a predator.

School A large number of fish swimming together.

Tentacles Long thin parts of an animal used to feel or grasp.

Transparent Clear and easy to see through, like glass.

Vulnerable Liable to be hurt.

Books to read

Bennett, Paul. *Catching a Meal*. New York: Thomson Learning, 1994.

Powzyk, Joyce. *Animal Camouflage: A Closer Look*. New York: Macmillan, 1990.

Sowler, Sandie. *Amazing Animal Disguises*. New York: Dorling Kindersley, 1992

Projects

Project: **Backyard (or local park)**

The creatures that live in your backyard face a constant battle for survival. They have many ways of escaping from their enemies. See if you can find examples of at least one animal for each of the methods of escape described in this book.

Here are some examples that may help you. Birds that visit your garden will fly away at the sight of a cat. Pill bugs will roll up into a ball to escape danger. Grasshoppers and many caterpillars are camouflaged to blend in with their surroundings. Hover-flies have stripes like a wasp, but do not sting. Ants live in colonies underground so that they can benefit from safety in numbers. Keep a record of all the examples that you find. Try and find out what their enemies are.

Project: **Animal Kingdom**

You can see many of the animals from this book in a well-maintained zoo, safari park, or public aquarium. See if you can find examples of at least one animal for each of the methods of escape described in this book. For example; you may find a porcupine, meerkat, stick insect, and tree frog in a zoo.

You may see a camouflaged flatfish, a shoal of fish, and a queen scallop in a public aquarium. Record all of your observations. It is impossible to show you all the examples of escape in just one book. See if you can find examples at the zoo/aquarium that are not in this book. What methods of escape do they use?

Index